This notebook belongs to:

AF882646

Proportions:

A **proportion** is when two ratios (or fractions) are equal.
We can use proportions to solve problems involving percents.

$$\frac{\text{part}}{\text{whole}} = \frac{\text{percent}}{100}$$

What is 25% of 160? $\frac{part}{160} = \frac{25}{100}$

Cross multiply: $100 \text{ X } part = 25 \times 160$

Divide: $part = \frac{160 \times 25}{100} = \frac{4000}{100} = 40$

The sale price of a phone was $140, which was 80% of the original price. What was the original price? ($175.)

Show how to find the answer:

Complete the Table of Commonly Used Values

Percent	Decimal	Fraction	Percent	Decimal	Fraction
1%				0.4	
		$^1/_{20}$		0.5	
10%			60%		
		$^1/_8$		0.75	
		$^1/_5$	80%		
		$^1/_4$			$^9/_{10}$
30%			99%		
		$^1/_3$			$^1/_1$

Rules of Exponents		Examples
Product Rule:	$a^m \cdot a^n = a^{m+n}$	$x^3 \cdot x^4 = x^{3+4} = x^7$
Power of a Product:	$(ab)^m = a^m b^m$	$(2y)^3 = 2^3 y^3 = 8y^3$
Power of a Power Rule:	$(a^m)^n = a^{m \cdot n}$	$(z^4)^3 = z^{4 \cdot 3} = z^{12}$
Quotient Rule:	$\frac{a^m}{a^n} = a^{m-n}$	$\frac{x^7}{x^4} = x^{7-4} = x^3$
Negative Exponent:	$a^{-m} = \frac{1}{a^m}$	$y^{-5} = \frac{1}{y^5}$
Zero Exponent:	$a^0 = 1$	$z^0 = 1$
Fractional Exponent:	$a^{\frac{x}{y}} = \sqrt[y]{a^x}$	$x^{\frac{2}{3}} = \sqrt[3]{x^2}$

Exponents and Fractions

$$(3x^{-3}y^2)^{-3} = 3^{-3}x^9y^{-6} = \frac{x^9}{3^3y^6} = \frac{x^9}{27y^6}$$

$$\frac{2^{-2}x^5y^3}{x^4y^6} = \frac{x^{5-4}}{2^2y^{6-3}} = \frac{x}{4y^3}$$

Cover the examples and solve.

$(3x^{-3}y^2)^{-3} =$

$\frac{2^{-2}x^5y^3}{x^4y^6} =$

Made in the USA
Monee, IL
07 November 2026

67873759R20079